Practice Exam for the Civil PE Examination

Breadth + Water Resources Depth

Indranil Goswami, Ph.D., P.E.

September 2015

Second printing September 2015

CIVIL PE WATER RESOURCES DEPTH EXAM – QUESTIONS © Copyright Indranil Goswami 2015

Preface

In January 2015, the official (NCEES) syllabus for the PE-Civil examination underwent a significant realignment. There was a significant departure from the previous structure which placed approximately equal emphasis on the five areas of practice. In the breadth exam, the 40 questions were approximately equally distributed among Construction, Geotechnical, Structural, Transportation and Water Resources. In the current syllabus for the Breadth (AM) exam, Transportation has been significantly deemphasized while there seems to be more emphasis on Construction.

The new depth (PM) syllabi have also gone through reorganization as well as addition of specific subtopics under various categories.

These practice exams were developed *after* the syllabus went through the aforementioned reorganization and are therefore consistent with the same.

This full-length practice exam contains 40 breadth (AM) questions + 40 depth (PM) questions in the area of CONSTRUCTION ENGINEERING. It should be taken under as near exam conditions as possible, preferably at the point when you think your exam review is complete and you are ready to take a simulated test to assess the level of your preparation. You should even go so far as to ask someone else to detach the questions from the solutions, so that you don't have any temptation to peek.

All the best for the upcoming PE exam,

Indranil Goswami

P.S. In this second printing, errors discovered to date have been corrected.

Table of Contents

BREADTH EXAM QUESTIONS 001 - 040	05-27
WATER RESOURCES DEPTH QUESTIONS 501 - 540	29-44
BREADTH EXAM ANSWER KEY	46
BREADTH EXAM SOLUTIONS 001 - 040	45-57
WATER RESOURCES DEPTH ANSWER KEY	60
WATER RESOURCES DEPTH SOLUTIONS 501 - 540	59-71

BREADTH EXAM
FOR THE
CIVIL PE EXAM

The following set of 40 questions (numbered 001 to 040) is representative of a 4-hour breadth (AM) exam according to the syllabus and guidelines for the Principles & Practice (P&P) of Civil Engineering Examination (updated January 2015) administered by the National Council of Examiners for Engineering and Surveying (NCEES). The exam is weighted according to the official NCEES syllabus (2015) in the following subject areas – Construction, Geotechnical, Structural, Transportation and Water & Environmental. Copyright and other intellectual property laws protect these materials. Reproduction or retransmission of the materials, in whole or in part, in any manner, without the prior written consent of the copyright holder, is a violation of copyright law.

The time allocated for this set of questions is 4 hours.

001

The tables below show historical data on traffic counts for a bridge, averaged by day of the week and month. If a daily count, conducted on a Wednesday in April is 19,545, the AAWT (Average Annual Weekday Traffic) for planning purposes is most nearly

    A. 16,300
    B. 17,250
    C. 22,830
    D. 23,440

Day of the week	ADT
Sunday	12,760
Monday	18,985
Tuesday	20,765
Wednesday	19,882
Thursday	20,349
Friday	16,889
Saturday	13,725
TOTAL	123,355

Month	ADT
January	17,756
February	16,772
March	19,674
April	21,983
May	20,935
June	16,783
July	15,887
August	16,785
September	19,836
October	19,356
November	20,128
December	19,785
TOTAL	225,680

002

The boundaries of a site form a triangle as shown below. The area of the site (acres) is most nearly

    A. 16
    B. 24
    C. 32
    D. 48

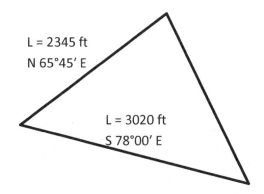

L = 2345 ft, N 65°45' E
L = 3020 ft, S 78°00' E

003

A reinforced concrete pipe of 36 inch outer diameter connects two manholes MH-1 and MH-2 as shown in the figure below. At station 13 + 05.10, the ground surface has a low point elevation of 242.35 ft. At this location, the soil cover (feet) is most nearly:

    A. 3.95
    B. 4.12
    C. 4.56
    D. 4.72

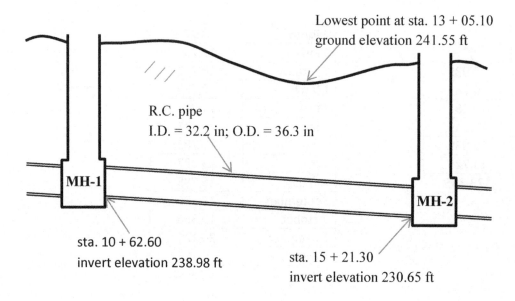

004

A roadside gutter is in the shape of a symmetric v-channel with 3H:1V sideslopes. The gutter is to be lined with a 3 inch thick concrete liner as shown. If the concrete material + placement cost is $232/yd^3, then the cost of constructing the gutters ($/mile) is most nearly

    A. 36,000
    B. 41,000
    C. 72,000
    D. 82,000

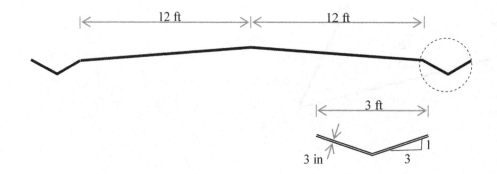

005

The activity on node network for a project is as shown below. All relationships are finish to start unless otherwise indicated. The table on the right shows pertinent data. The early start date (weeks) for activity F is:

A. 15
B. 16
C. 17
D. 18

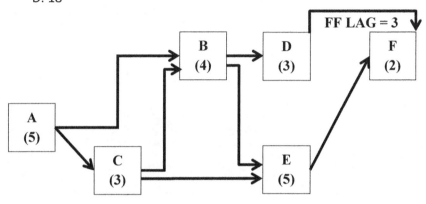

006

A parabolic vertical curve must connect a tangent of slope +5% to another of slope -3% as shown below. The two tangents intersect at a point at station 11 + 45.20 and at elevation 310.56 ft. A sewer (circular CIP) with crown elevation 302.65 ft exists at station 12 + 30.05. If minimum soil cover of 30 inches is required above the sewer pipe, the required length of curve (feet) is most nearly

A. 510
B. 570
C. 635
D. 815

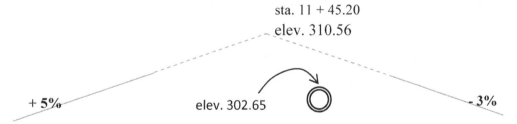

007

A cantilever retaining wall is supported by a 3-foot thick footing as shown. The drains behind the wall become clogged and groundwater rises to the top of the horizontal backfill. The total horizontal earth pressure resultant (lb/ft) acting on the retaining wall is most nearly:

    A. 22,300
    B. 20,650
    C. 18,770
    D. 13,300

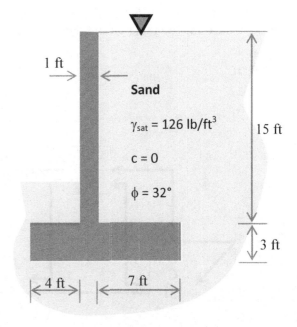

008

A 10-ft layer of varved clay is overlain by a 12 ft thick sand layer, as shown. The water table is originally at a depth of 5 ft below the ground surface. Prior to construction, the water table is lowered by 7 ft, to the bottom of the sand layer. Three months after lowering the water table, settlement (inches) due to consolidation of the clay layer is most nearly

    A. 0.9
    B. 1.9
    C. 2.6
    D. 4.0

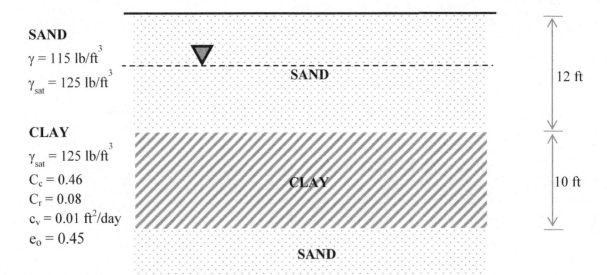

CIVIL PE WATER RESOURCES DEPTH EXAM – QUESTIONS © Copyright Indranil Goswami 2015

009

Identify the correct shape of the bending moment diagram for the beam loaded as shown below.

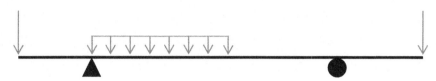

A.

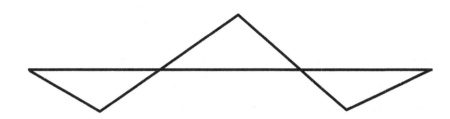

B.

C.

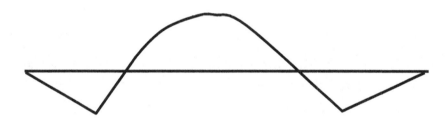

D.

010

A rectangular post is subject to an eccentric load P as shown. The maximum compressive stress (MPa) is most nearly

    A. 1.0
    B. 1.6
    C. 2.6
    D. 4.0

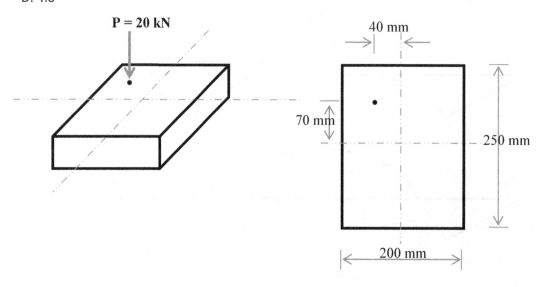

011

A trapezoidal open channel conveys flow at a uniform depth of 5 ft as shown below. The Manning's n = 0.015. The bottom width of the channel = 20 ft and longitudinal slope of the channel floor is 0.8%.

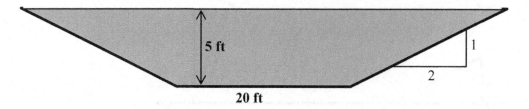

The flow rate (MGD) conveyed by the channel is most nearly:

    A. 500
    B. 1000
    C. 2000
    D. 3000

012

A 5 ft x 5 ft square footing transfers a column load of 140 kips to a sandy soil as shown. The depth of the footing is 3 ft. The factor of safety against general bearing capacity failure is most nearly

   A. 3.3
   B. 2.7
   C. 2.0
   D. 1.3

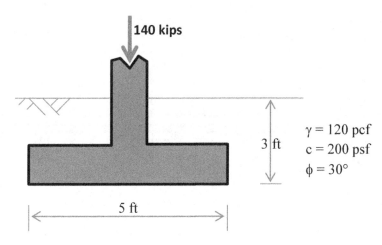

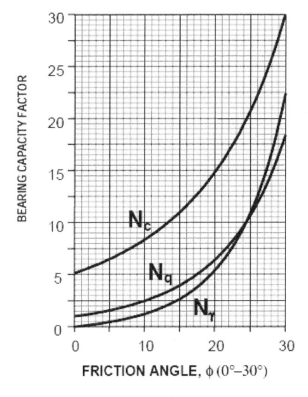

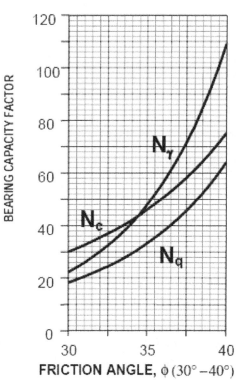

013

A wall panel is composed of plywood sheathing supported by 2x6 studs (nominal dimensions 1.5 in x 5.5 in) spaced every 28 inches as shown. The studs are supported by longitudinal members every 10 ft. The wall experiences a normal wind pressure of 30 psf. The maximum bending stress (lb/in^2) in the studs is most nearly:

A. 800
B. 1000
C. 1200
D. 1400

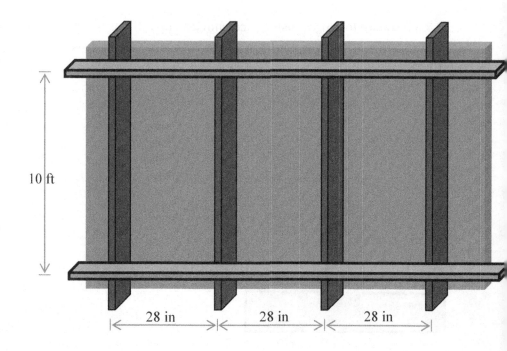

014

The table below shows cross section areas of cut and fill recorded at 5 stations spaced at 100 ft.

Station	Area (ft^2)	
	CUT	FILL
0 + 0.00	245.0	123.5
1 + 0.00	312.5	76.3
2 + 0.00	411.5	0.0
3 + 0.00	234.5	88.4
4 + 0.00	546.2	214.5

The net earthwork volume (yd^3) between stations 0 + 0.00 and 4 + 0.00 is most nearly

A. 3640 (cut)
B. 3780 (cut)
C. 3640 (fill)
D. 3780 (fill)

015

A site needs soil compacted to 90% of the Proctor maximum dry density. The results of the Proctor test are shown below.

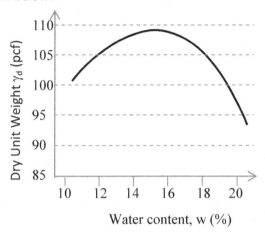

The volume of the embankment is 50,000 ft³. If borrow soil is available at γ = 120 pcf and moisture content = 14%, the volume of soil (yd³) needed from the borrow pit is most nearly:

    A. 1,400
    B. 1,730
    C. 2,080
    D. 2,800

016

Isohyets showing precipitation depth are shown in the figure. The accompanying table shows total area enclosed by each close contour. The average precipitation depth (inches) is most nearly:

    A. 0.46
    B. 0.51
    C. 0.58
    D. 0.62

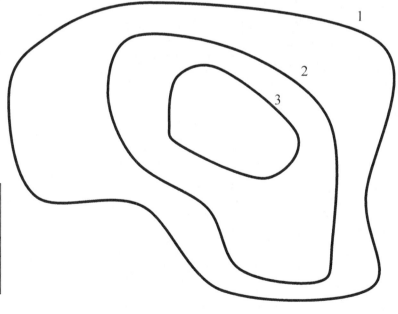

Contour	Area enclosed (acres)	Precip. depth (inches)
1	268	0.2
2	121	0.6
3	45	0.9

017

A 24 inch diameter reinforced concrete pipe (C = 100) conveys water a flow rate = 12.5 ft³/sec. The head loss due to friction (feet per mile) is most nearly:

   A. 11.5
   B. 14.6
   C. 16.3
   D. 18.3

018

Runoff flow from a development is held in a detention pond until it is 78% full, at which point, it empties through a weir. The capacity of the pond is 760,000 gallons. If after a rainfall event, the average rate of inflow into the pond occurs at 2 ft³/sec, the length of time (hours) before the pond starts to empty is most nearly:

   A. 7.5
   B. 9.3
   C. 11.0
   D. 13.7

019

For the truss shown below, the axial force (kips) in member CD is most nearly:

   A. 125 (tension)
   B. 125 (compression)
   C. 150 (tension)
   D. 150 (compression)

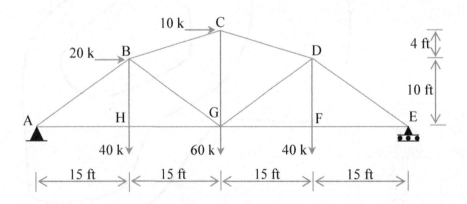

020

A 6-inch thick riprap layer is used as protection for the earth slope (θ = 30°) shown below. The factor of safety for slope stability is most nearly:

    A. 1.25
    B. 1.35
    C. 1.45
    D. 1.55

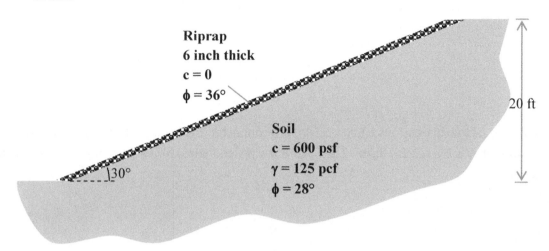

021

Particle size distribution of a soil sample is summarized in the table below. The fines were tested using Atterberg apparatus to obtain the following results:

Liquid limit   45
Plastic limit   21

Sieve size	% finer
1.0 inch (25.4 mm)	100
0.5 inch (12.7 mm)	92
No. 4 (4.75 mm)	75
No. 10 (2.00 mm)	62
No. 40 (0.425 mm)	52
No. 100 (0.15 mm)	45
No. 200 (0.075 mm)	28

The classification of the soil according to the USCS is:

    A. GW
    B. SM
    C. SC
    D. GC

022

A soil sample yields the following results:
Mass of wet soil = 1685 g
Volume of wet soil = 855 cc
Mass of soil after oven drying = 1418 g
If the specific gravity of soil solids is taken as 2.65, the void ratio is most nearly:
    A. 0.3
    B. 0.4
    C. 0.5
    D. 0.6

023

2x6 stud columns (actual dimensions 1.5 in x 5.5 in) are connected to plywood sheathing as shown. The modulus of elasticity of timber is E = 1.5×10^6 lb/in^2. The Euler buckling load (kips) for each column is most nearly:
    A. 9
    B. 15
    C. 21
    D. 26

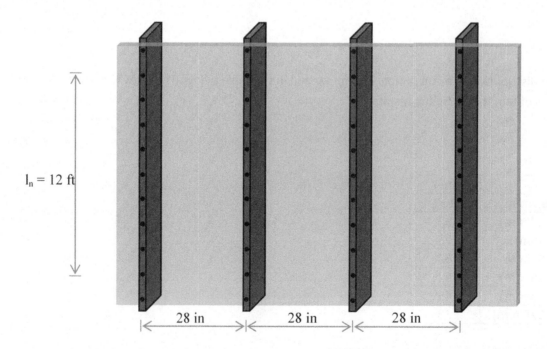

024

Water stored in a large reservoir (surface elevation 324.5 ft above sea level) empties through a 24 inch diameter pipe as shown. The far end of the pipe is at elevation 295.8 ft above sea level. The discharge (ft³/sec) through the pipe is most nearly:

   A. 50
   B. 38
   C. 32
   D. 22

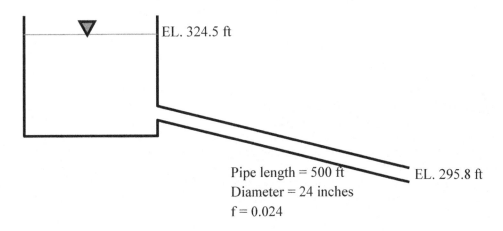

025

Surface runoff from a catchment area has the longest hydraulic path as shown in the figure below. Time for sheet flow, t_s = 5 minutes and for ditch flow t_d = 13 minutes. A collector pipe, in which flow occurs at an average velocity of 5 feet/sec, of length 1200 ft has an inlet at point A and discharges into a main sewer at point B. Intensity-duration-frequency curves are obtained from historical precipitation data. The design intensity (in/hr) to be used for the design of the sewer mains for a 20 year storm is most nearly:

   A. 1.5
   B. 1.8
   C. 2.1
   D. 2.4

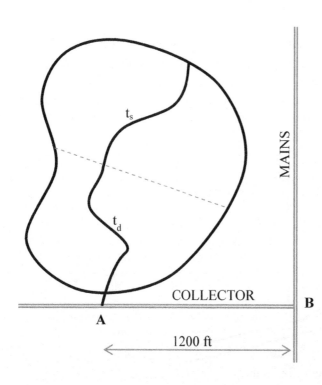

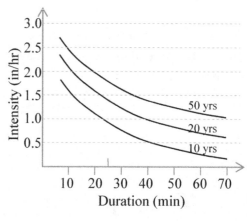

026

A circular horizontal curve has the coordinates (feet) of the PC, PI and PT as shown on the figure.

The degree of curve (degrees) is most nearly:

    A. 10
    B. 13
    C. 16
    D. 19

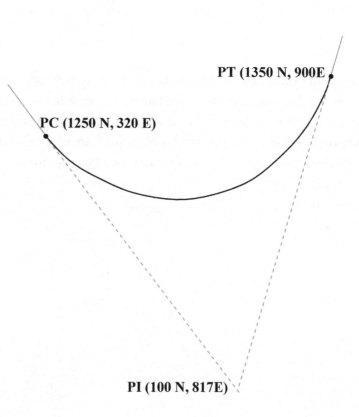

027

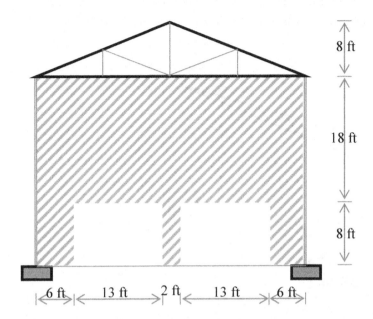

A 110 ft long warehouse shed has metal sheathing forming the sides (no openings) and front and back (with two door openings on each surface) and the roof. The total surface area of metal sheathing (sq. ft.) is most nearly:

A. 12,652
B. 12,456
C. 12,331
D. 12,123

028

A crane with a 40 ft boom is used to lift a 4 ton load as shown. The total weight of the crane and ballast is 4.5 tons acting at the effective location indicated as CG on the figure. The weight of the boom is 800 lb. Each outrigger leg is supported by a circular pad with a diameter = 3 feet

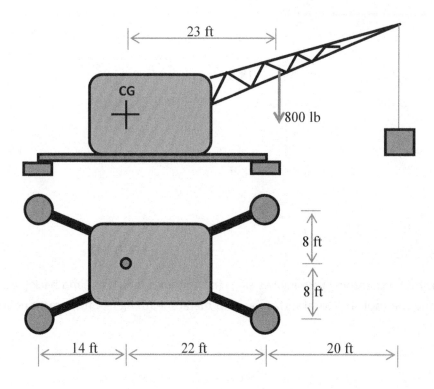

The maximum soil pressure (lb/ft²) under the outrigger pads is most nearly:

    A. 1,200
    B. 1,600
    C. 2,000
    D. 2,400

029

Which of the following statements is true about deflection of concrete beams?
I.   The deflection is calculated using the moment of inertia of the uncracked section
II. The deflection is calculated using a moment of inertia equal to half the gross moment of inertia
III. The deflection is calculated using the moment of inertia of the cracked section
IV. The deflection is calculated using a moment of inertia less than the gross moment of inertia

    A. I and IV
    B. IV only
    C. II only
    D. II and III

030

The figure shows a stress-strain diagram based on a tension test of a steel test coupon.

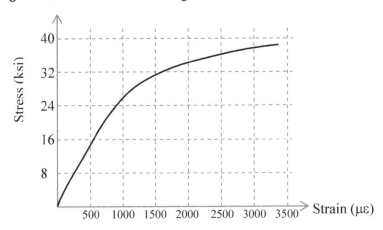

The yield stress (kip/in²) is most nearly
- A. 26
- B. 30
- C. 34
- D. 38

031

A layer of coarse sand (thickness = 15 feet) supports a mat foundation that exerts a net uniform pressure of 600 psf at a depth of 3 feet below the surface as shown. The compression of the sand layer (inches) is most nearly:
- A. 0.1
- B. 0.2
- C. 0.3
- D. 0.4

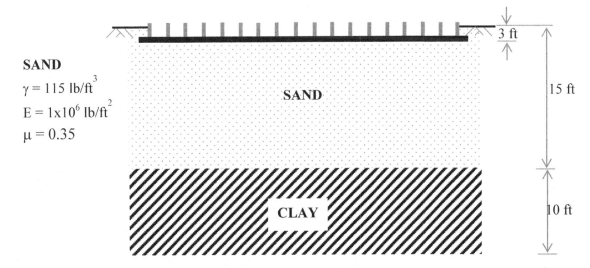

032

Which of the following statements about total float (TF) in a CPM network is not true? Use the following nomenclature – ES (early start), EF (late finish), LS (late start), LF (late finish), D (duration).
I.   TF = LF – ES – D
II. TF = LS – EF – D
III. TF = LS – EF + D
IV. TF = LS – ES
   A. I
   B. II
   C. III
   D. IV

033

Which of the following statements is/are true?
I.   For long term stability analysis of clay slopes, results of the CD triaxial test must be used.
II. The CD triaxial test takes longer to perform than the UU test.
III. The UU triaxial test takes longer to perform than the CD test
IV. Pore pressure measurements must be made during the CD test

   A. I, II only
   B. II, III only
   C. III only
   D. all of them

034

A test strip shows that a steel-wheeler roller, operating at 3 mph, can compact a 0.5 ft. layer of material to a proper density in four passes. The width of the drum is 8.0 ft. The roller operates 50 min per hour. The number of rollers required to keep up with a material delivery rate of 540 bank cubic yards/hr is most nearly: (1 bank cubic yard = 0.83 compacted cubic yard):
   A. 4
   B. 3
   C. 2
   D. 1

035

A simply supported steel beam carries a single concentrated load at midspan as shown. The beam as the following properties: Area A = 29.4 in^2; I$_x$ = 1490 in^4; I$_y$ = 186 in^4; Z$_x$ = 198 in^3; Z$_y$ = 54.9 in^3; S$_x$ = 175 in^3; S$_y$ 35.7 = in^3.

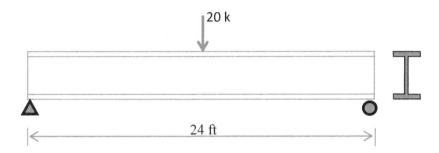

The maximum deflection (inches) is most nearly:

    A. 0.25
    B. 0.50
    C. 0.75
    D. 0.90

036

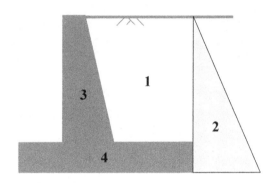

	Component	Resultant Force
1	Backfill soil	12,600 lb/ft
2	Active earth pressure	5,400 lb/ft
3	Concrete wall stem	3,500 lb/ft
4	Concrete wall footing	4,500 lb/ft

A concrete retaining wall has a level backfill behind it. The coefficient of friction between the wall footing and the soil is 0.6. The table to the right summarizes the forces acting on the retaining wall. The factor of safety against sliding is most nearly:

    A. 0.9
    B. 1.6
    C. 2.3
    D. 2.9

037

The 'first flush' runoff depth (assumed to be 1 inch) from a 120 acre watershed collects in a 2 acre detention pond. The sediment load carried by the runoff is 5 g/L. Bulk density of sediment is 80 lb/ft³. The loss of depth (inches) per rainfall event is most nearly:

A. 0.125
B. 0.25
C. 0.50
D. 0.75

038

Several cylindrical steel samples are tested in a Universal Testing Machine to obtain the results obtained below

Sample	Diameter (in)	Sample length (in)	Breaking load (lb)	Elongation at Failure (in)
1	0.504	5.66	9050	0.412
2	0.498	7.34	8865	0.516
3	0.509	7.55	9235	0.543
4	0.503	8.12	9110	0.581
5	0.512	7.12	9565	0.505

The average breaking strain ($\mu\varepsilon$) is most nearly:

A. 38,500
B. 51,200
C. 71,500
D. 78,300

039

A circular conduit of diameter 48 inches conveys water at a depth of 30 inches as shown below. The interior of the concrete pipe is coated to yield n = 0.014 (assumed constant with varying depth).

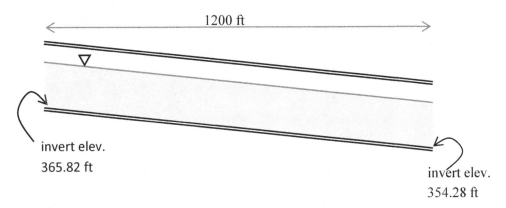

The velocity (ft/sec) is most nearly:
- A. 8.2
- B. 9.3
- C. 10.4
- D. 11.3

040

Which of the following techniques are commonly used for construction adjacent to historic structures?
- I. Underpinning
- II. Anchor rod and deadman
- III. Slurry walls
- IV. Compaction piles

A. I and II
B. I and III
C. II and III
D. II and IV

THIS IS THE END OF THE BREADTH EXAM

WATER & ENVIRONMENTAL DEPTH EXAM

FOR THE

CIVIL PE EXAM

The following set of 40 questions (numbered 501 to 540) is representative of a 4-hour depth (PM) exam for WATER & ENVIRONMENTAL according to the syllabus and guidelines for the Principles & Practice (P&P) of Civil Engineering Examination (updated January 2015) administered by the National Council of Examiners for Engineering and Surveying (NCEES). Copyright and other intellectual property laws protect these materials. Reproduction or retransmission of the materials, in whole or in part, in any manner, without the prior written consent of the copyright holder, is a violation of copyright law.

The time allocated for this set of questions is 4 hours.

The schematic of a wastewater treatment plant is show below. Questions 501 – 503 are based on this schematic.

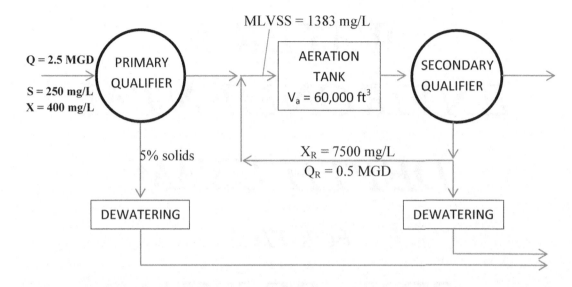

501

The TSS for the influent into the primary clarifier is 400 mg/L. The primary clarifier reduces total suspended solids by 60% and BOD by 15%. The volume of primary sludge produced per day (gal/day) is most nearly:

    A. 125,000
    B. 65,000
    C. 12,000
    D. 8,500

502

The hydraulic detention time for the aeration tank (hours) is most nearly:

    A. 3.6
    B. 4.8
    C. 7.2
    D. 9.6

503

The food to microorganism ratio for the aeration tank (day^{-1}) is most nearly:

    A. 0.1
    B. 0.5
    C. 0.7
    D. 1.0

504

A Pitot tube is used to measure the rate of flow in a conduit as shown below. The height of the water column in the tube is 22 inches.

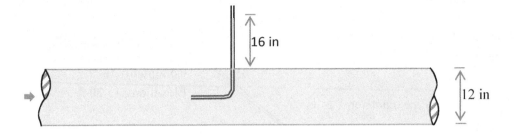

The pressure in the conduit (psig) is most nearly:
- A. 0.2
- B. 0.8
- C. 14.9
- D. 15.5

505

A water delivery pipeline has a diameter of 12 inches. The elevation of the energy grade line at the upstream end (A) is 239.89 ft above sea level. The total head loss over the length AB of the pipe is equivalent to 28.7 ft. The flow in the pipe empties through an open end at B, which is at elevation 200.50 ft.

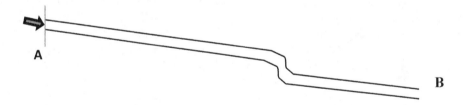

The flow rate in the pipe (gal/min) is most nearly:
- A. 7,670
- B. 9,250
- C. 12,150
- D. 14,255

506

A centrifugal pump, operating at 88% efficiency, is used to pump water from reservoir A to a community at B as shown. The total head loss in the delivery system is estimated to be 45 ft. The total flow rate is 9000 gal/min and the desired gage pressure at the community is 60 psi.

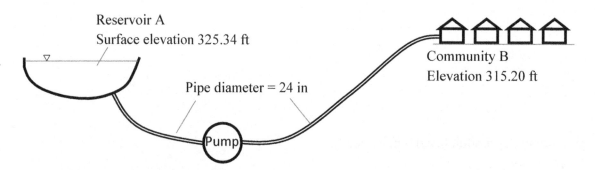

The required power of the pump (hp) is most nearly:

    A. 300
    B. 350
    C. 400
    D. 450

507

A pipe network is shown below. Physical data for each pipe segment are given in the table.

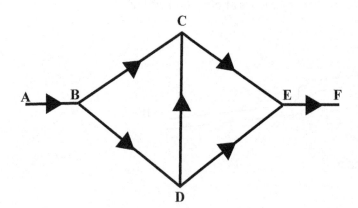

Pipe	Length (ft)	Dia (in)	Friction factor (f)	Flow rate (gpm)
AB	200	12	0.020	900
BC	350	8	0.030	465
BD	450	8	0.024	435
CD	600	8	0.025	64
CE	800	12	0.030	529
DE	500	8	0.016	371
EF	300	12	0.020	900

The pressure loss (lb/in^2) between A and F is most nearly:

    A. 1.7
    B. 2.5
    C. 3.6
    D. 5.3

508

Oil (S.G. = 0.9) flows through a short reducer (d_1 = 6 inches, d_2 = 4 inches) at a flow rate Q = 3.0 ft^3/sec. If the energy losses are negligible, the pressure loss (lb/in^2) in the reducer is most nearly:

    A. 8.8
    B. 6.5
    C. 5.8
    D. 3.4

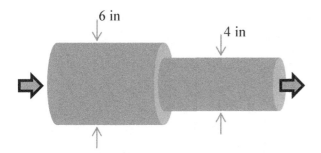

509

A reinforced concrete pipe (diameter 18 inches) conveys a flow rate = 1500 gal/min. The Hazen Williams roughness coefficient of the pipe is 110. The slope of the energy grade line (ft/ft) is most nearly:

    A. 0.001
    B. 0.002
    C. 0.003
    D. 0.004

510

A trapezoidal open channel conveys flow at a depth of 4 ft as shown below. The Manning's n = 0.015. The bottom width of the channel = 12 ft and longitudinal slope of the channel floor is 0.8%. The flow rate is 405 ft^3/sec.

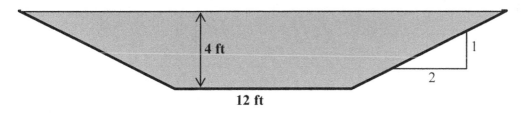

The Froude number is most nearly:

    A. 0.21
    B. 0.41
    C. 0.53
    D. 0.88

511

A trapezoidal open channel conveys a flow rate = 400,000 gal/min. The Manning's n = 0.015. The bottom width of the channel = 20 ft and longitudinal slope of the channel floor is 0.8%. The side slopes are 2:1

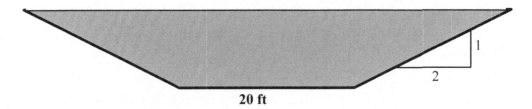

The depth of flow (feet) is most nearly:
  A. 5.00
  B. 4.25
  C. 3.75
  D. 2.50

512

Which of the following is/are effective as an energy dissipating device for flow in open channels?
  I. hydraulic jump
  II. sudden expansion
  III. floor roughening
  IV. floor raising
  V. baffles
    A. I, II and III
    B. I and V
    C. All of the above
    D. I, II and V

513

Surface runoff collects from a watershed into a channel which flows from west to east as shown. Inlets 1, 2 and 3 collect overland flow from regions A, B and C respectively. The average velocity of flow in the channel can be assumed to be 4 feet per second.

The time of concentration (minutes) for point X

A. 65
B. 63
C. 59
D. 56

Region	Area (acres)	Rational C	Time for overland flow (min)
A	80	0.45	45
B	95	0.35	32
C	135	0.25	56

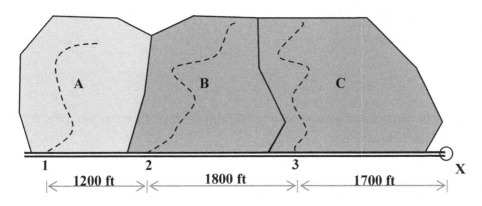

514

Water flows in a 20 ft wide rectangular open channel at a supercritical depth of 12 inches. The flow then goes through a hydraulic jump as shown in the figure. The fraction of incident energy that is lost through the hydraulic jump is most nearly:

A. 64%
B. 54%
C. 44%
D. 34%

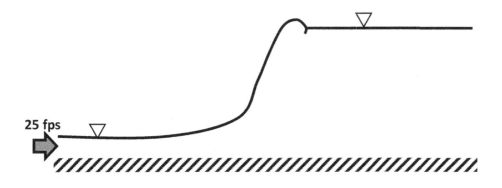

515

A community is subdivided into 4 zones of land use and soil cover as summarized in the table below. For a 2 hr storm with a gross rainfall depth of 3.2 inches, the runoff depth (inches) predicted by the NRCS method is most nearly:

   A. 0.9
   B. 1.2
   C. 1.5
   D. 1.8

Region	Description	Area (acres)	CN	Rational C
A	Well maintained lawns	25	74	0.70
B	Paved (driveways etc.)	5	98	0.95
C	Single family homes	12	80	0.75
D	Forested	34	64	0.60

516

The average infiltration rate for buried water mains is given as 100 gpd per inch diameter per mile. A water distribution network consists of the following pipe inventory:

Diameter (in)	Length (feet)
8	13,400
12	7,500
20	4,000
36	3,000

The extraneous flow rate (ft^3/sec) due to infiltration is most nearly:

   A. 0.034
   B. 0.023
   C. 0.017
   D. 0.011

517

According to the U.S. Soil Conservation Service, the number of rainfall distribution types recognized in the United States is

   A. 3
   B. 4
   C. 5
   D. 6

518

A stream collects runoff from a watershed area A = 325 acres. The table below shows discharge (ft³/sec) recorded in the stream during the period t = 0 to t = 5 hours after a rainfall event.

Time (hr)	Flow Rate (ft³/sec)
0	20
1	50
2	70
3	120
4	60
5	20

The average depth of runoff (inches) from the watershed is most nearly:

A. 0.36
B. 0.54
C. 0.67
D. 1.04

519

A chlorinator is set for a feed rate 50 lbs of chlorine per 24 hours and the wastewater flow is 1.1 MGD. After 15 minutes of contact, the chlorine is measured as 0.5 mg/L. The chlorine demand (mg/L) is most nearly:

A. 3.5
B. 4.0
C. 4.5
D. 5.0

520

An aerobic digester is 100 ft long by 40 ft wide by 12 ft deep. The sludge contains 2% solids. The digester is decanted at the rate of 2000 gal/day. The decanted liquid contains solids concentration of 400 mg/L. Solids are removed from the digester at the rate of 4000 gal/day. The solids retention time (days) for the digester is most nearly

A. 90
B. 70
C. 60
D. 35

521

An industrial plant discharges waste water at a rate of 4000 gal/min. The discharge is contaminated with copper concentration of 4 mg/L. The wastewater goes through tertiary treatment for the removal of copper and is then discharged into a stream which has an upstream flow rate = 300 ft³/sec. The water in the stream carries negligible copper. If the state standards require that copper concentration in the stream never exceed 20 µg/L, the removal rate (%) required for the tertiary process is most nearly

A. 53
B. 75
C. 85
D. 97

522

A small township has 2785 households and a population of 12,300. The wastewater flow rate (MGD) during the AM peak hour is most nearly:

    A. 1.4
    B. 2.9
    C. 4.0
    D. 5.2

523

Water gets pumped at the rate of 800 gpm from a 32 ft thick confined aquifer. The bottom of the aquifer is at a rocky layer at elevation 108.78 ft above sea level and the elevation of the piezometric surface is 205.98 ft above sea level before the pumping begins. The radius of influence of the well is 1250 ft.

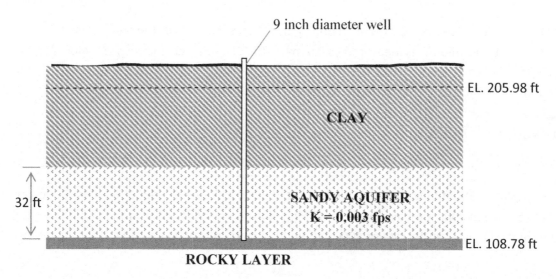

Once the drawdown reaches steady state, the maximum drawdown (feet) of the piezometric surface is most nearly:

    A. 194.5
    B. 197.7
    C. 199.2
    D. 201.6

524

Flash runoff of 1 inch depth from a 60 acre community flows into a detention pond with surface area 1.4 acre. Evaporation from the surface of the pond is 2 gpd/ft^2. A discharge weir releases water at an average rate of 1000 gpm. Elevation of the water surface in the pond before the flash flood is 230 ft above sea level. 1 hour after the runoff enters the pond, the surface elevation (feet above sea level) is most nearly:

    A. 228.15
    B. 231.15
    C. 233.15
    D. 235.15

525

A sample of municipal water has the following ion concentrations:

Ca^{2+}	15 mg/L
Mg^{2+}	20 mg/L
Na^+	22 mg/L
Fe^{3+}	8 mg/L
OH^-	35 mg/L
SO_4^{2-}	15 mg/L
HCO_3^-	25 mg/L
CO_3^{2-}	30 mg/L

The total hardness (mg/L as $CaCO_3$) is most nearly:

    A. 43

    B. 65

    C. 190

    D. 140

526

A sedimentation tank receives 3 MGD of wastewater containing TSS = 380 mg/L. Suspended solids are 60% volatile. The reactor achieves 80% removal of FSS and negligible organic solids removal. The amount of solids (lb) removed per day is most nearly:

    A. 3,000

    B. 3,700

    C. 4,500

    D. 5,200

527

A drinking water sample is analyzed with the following (partial) results:

Sodium [Na^+]	1.8 mg/L
Total dissolved solids	255 mg/L
Turbidity	1.7 NTU
Odor	2.3 TON
Nitrite [NO_2^-]	3 mg/L
Fluoride [F^-]	2 mg/L

Which of the following impurities are in violation of EPA's drinking water standards?

    A. Turbidity only

    B. Nitrite, odor and TDS

    C. Turbidity, TDS and fluoride

    D. Turbidity and odor

528

Dissolved oxygen concentration in a stream changes continuously as a result of deoxygenation (caused by organic load carried by wastewater) and reoxygenation (caused by the interaction of the stream and the atmospheric air). An empirical model suggested by the USGS suggests:

$$k_r = \frac{3.3V}{H^{1.33}}$$

where k_r = reaeration coefficient (day^{-1})
 V = average stream velocity (ft/sec)
 H = stream depth (ft)

For a natural stream with the following parameters, the reaeration coefficient (day^{-1}) is most nearly:
 Flow rate = 2,514 ft^3/sec
 Area of flow section = 1323 ft^2
 Wetted perimeter = 517 ft
 Average depth of flow = 3.8 ft
 A. 0.3
 B. 0.5
 C. 0.8
 D. 1.1

529

The flow net shown below describes the seepage into a long 20 feet deep and 40 feet wide excavation made in silty sand having a coefficient of permeability (K) equal to 3 x 10^{-4} cm/sec.

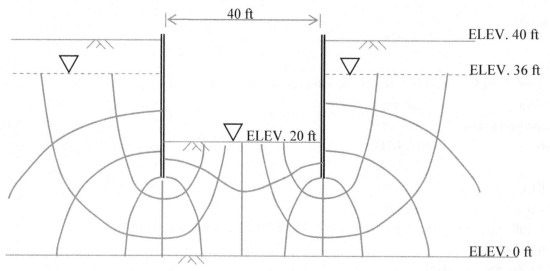

To maintain the water level at the base of the excavation the quantity of water the contractor has to pump is most nearly:
 A. 3.0 x 10^{-3} ft^3/min/ft
 B. 8.0 x 10^{-3} ft^3/min/ft
 C. 15.0 x 10^{-3} ft^3/min/ft
 D. 30.0 x 10^{-3} ft^3/min/ft

530

The 1 hour unit hydrograph shown in the table below is used to predict the runoff produced by storms. The storm in question is one that produces 0.5 in of excess precipitation during the first hour, 0.9 inches during the second hour and 0.3 inches during the third hour.

Time (hr)	Flow Rate (ft^3/sec/in)
0	0
1	80
2	180
3	320
4	210
5	0

The peak discharge (ft^3/sec) is most nearly:
- A. 250
- B. 350
- C. 450
- D. 550

531

A concrete sewer of diameter 48 inches conveys flow at a depth of 30 inches. The longitudinal slope of the sewer is 1% and Manning's n = 0.015.

The flow rate (MGD) is most nearly:
- A. 57
- B. 80
- C. 96
- D. 118

532

The flow rate treated at a wastewater treatment plant is 3 MGD. The concentration of total suspended solids (TSS) in the influent is 1,100 mg/L. The flow passes through a bank of filters, arranged in parallel. The maximum solids load on each filter is 12 lb-TSS/ft^2-day. Plant operation guidelines require at least two filters to be shut down at any time for backwashing. If each filter measures 15 ft x 15 ft, the number of filters needed is most nearly:
- A. 10
- B. 11
- C. 12
- D. 13

533

A rain gage at a weather station records the following rainfall depths during a storm event. The peak intensity of rainfall (in/hr) during the storm is most nearly:

A. 0.6
B. 2.0
C. 3.2
D. 3.6

Time (min)	Depth (inches)
0	0.0
10	0.2
20	0.5
30	0.9
40	1.5
50	1.8
60	2.0

534

A dam of length 120 ft is constructed of an impermeable soil. The dam overlays a sand bed of thickness 6.5 ft as shown. The permeability of the sand bed is 9.5 ft/day.

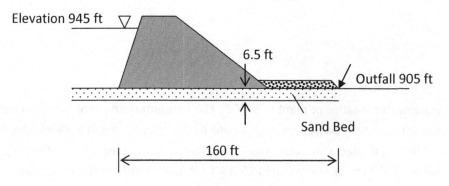

The discharge due to seepage (gal/min) is most nearly:

A. 7.5
B. 9.5
C. 11.5
D. 13.5

535

A standard BOD test conducted on a wastewater sample has the following data:

Sample volume = 15 mL
Volume of dilution water added = 285 mL
Initial (at t = 0) dissolved oxygen concentration = 6.3 mg/L
Final (at t = 5 days dissolved oxygen concentration = 2.5 mg/L
Incubation temperature = 25°C
Deoxygenation rate constant (base 10, 25°C) = 0.10

The ultimate BOD (mg/L) of the sample is most nearly:

A. 75
B. 110
C. 130
D. 150

536

Methylene chloride or dichloromethane is a widely used solvent. It poses a hazard via inhalation and absorption through the skin. The Reference Dose (R_fD) for methylene chloride is 0.06 mg per kg body weight per day (mg/kg/d) based on liver toxicity in rats.

For an adult with body weight 70 kg with a lifetime water consumption of 2 L/day, the drinking water equivalent level (mg/L) is most nearly:

A. 0.06
B. 0.2
C. 0.6
D. 2.0

537

A rapid mix tank receives a wastewater flow rate of 1.5 MGD. The water has a temperature of 50°F and pH = 6.7. Hydraulic detention time = 2 minutes. If the impeller power available for mixing is 35 hp, the velocity gradient (sec^{-1}) is most nearly:

A. 2,200
B. 1,600
C. 1,100
D. 800

538

Which of the following is an example of bioaccumulation?

    A. buildup of greenhouse gases in the atmosphere

    B. algae blooms in ponds

    C. buildup of arsenic in human tissue from contaminated water

    D. sedimentation of silts in detention ponds

539

Which of the following compounds is commonly used for taste and odor control of drinking water?

    A. Manganese sulfate

    B. Potassium permanganate

    C. Manganese dioxide

    D. Potassium dichromate

540

A structure has been examined to determine a need for rehabilitation. The related costs are summarized below:

    Current annual costs = $40,000

    Estimated rehabilitation cost = $350,000

    Annual costs projected after rehabilitation = $15,000

    Expected useful life remaining = 20 years

    Projected increase in residual value (at end of useful life) = $200,000

The return on investment (ROI) for performing the rehabilitation is most nearly:

    A. 5%

    B. 6%

    C. 7%

    D. 8%

THIS IS THE END OF THE WATER RESOURCES DEPTH EXAM

SOLUTIONS TO BREADTH EXAM
FOR THE
CIVIL PE EXAM

ANSWER KEY: BREADTH EXAM

001	A
002	D
003	B
004	C
005	C
006	B
007	D
008	A

009	C
010	B
011	C
012	A
013	D
014	B
015	B
016	C

017	D
018	C
019	B
020	A
021	C
022	D
023	B
024	A

025	A
026	D
027	D
028	A
029	B
030	D
031	A
032	B

033	A
034	D
035	A
036	C
037	B
038	C
039	D
040	B

Solution 001

From the table on the right, the total annual count = 225,680, from which we obtain an average monthly ADT = 18,807

Therefore, the monthly expansion factor for April = 18,807÷21,983 = 0.856

From the table on the left, the 7-day count = 123,355, of which 26,485 is weekend traffic. Therefore, the cumulative (Mon-Fri) weekday traffic is 96,870 (5 day average of 19,374.

Therefore, the daily expansion factor for Wednesday = 19,374÷19,882 = 0.974

Therefore the AAWT = 19,545x0.856x0.974 = 16,296

Answer is A

Solution 002

The easiest way to solve this problem is to use the formula for the area of a triangle:
$$A = \frac{1}{2}ab \sin C$$
where C is the angle between two sides a and b

The azimuth angles of the two lines are 65.75 and 102 degrees. Therefore the angle between them is 102 − 65.75 = 36.25

Area: $A = \frac{1}{2} \times 2345 \times 3020 \times \sin 36.25 = 2{,}093{,}798 \: ft^2 = 48.067 \: acres$

Answer is D

Alternatively: Arbitrarily assuming the coordinates of A to be (0, 0), the coordinates of the other two points can be found and then the method of coordinates can be used.

Solution 003

The pipe length between the two manholes has length = 1521.3 − 1062.6 = 458.70 ft.

The point of interest (low ground elevation) is located at a distance = 1305.1 – 1062.6 = 242.50 ft. from the upstream end, and 216.2 ft from the downstream end

The invert elevation at this location can be calculated by averaging the upstream and downstream invert elevations, as follows

$$238.98 \times \frac{216.2}{458.7} + 230.65 \times \frac{242.5}{458.7} = 234.58$$

Pipe wall thickness = (36.3 – 32.2)/2 = 2.05 in = 0.171 ft
Outer diameter = 36.3 in = 3.025
Elevation of the TOP of pipe = 234.58 – 0.171 + 3.025 = 237.43 ft
Soil cover = 241.55 – 237.43 = 4.12 ft

Answer is B

Solution 004

The perimeter of each gutter is 3.162 ft
Total perimeter of gutters = 6.324 ft
Surface area per mile = 6.324x5280 = 33,393.65 ft^2/mile
With a thickness of 3 inches, the volume of concrete = 8,348.4 ft^3/mile = 309.2 yd^3/mile
Cost of concrete material and placement = 232 x 309.2 = $71,734.40

Answer is C

Solution 005

Starting from A: ES_A = 0; EF_A = 0 + 5 = 5. This carries over to the successor C
For activity C: ES_C = 5; EF_C = 5 + 3 = 8
For activity B, there are two predecessors (A and C): ES_B = larger of EF_A and EF_C = 8; EF_B = 8 + 4 = 12
Since D has a single predecessor (B), ES_D = EF_B = 12. Therefore EF_D = 12 + 3 =15
E has two predecessors (B and C). Therefore ES_E = larger of EF_B and EF_C = 12. And EF_E = 12 + 5 = 17
Based on the FF lag between D and F, the EF_F = 18, based on which the ES_F = 16. However, based on EF_E = 17, the ES_F = 17. This controls.

Answer is C

Solution 006

The elevation of the point on the curve is 2.5 ft (30 in) above the crown of the sewer pipe, therefore at elev. 305.15. This point has the following offsets from the PVI:

horizontal offset h = 1230.05 – 1145.20 = +84.85 ft

vertical offset v = 305.15 – 310.56 = - 5.41 ft

The maximum length of curve (L) can be calculated from

$$\frac{L+2h}{L-2h} = \sqrt{\frac{v-G_1h}{v-G_2h}} = \sqrt{\frac{-5.41-0.05\times 84.85}{-5.41--0.03\times 84.85}} = \sqrt{\frac{-9.6525}{-2.8645}} = 1.836$$

$$\frac{L+169.7}{L-169.7} = 1.836$$

Solving L = 575.7 ft. Since this is the maximum length of curve, look for the next lower value

Answer is B

Solution 007

For ϕ = 32, active earth pressure coefficient: $K_a = \frac{1-\sin\phi}{1+\sin\phi} = 0.307$

At the base of the footing (depth = 18 ft), the effective earth pressure: $K_a\gamma_{sub}H = 0.307 \times (126-62.4) \times 18 = 351.5\ psf$

At the base of the footing (depth = 18 ft), the hydrostatic pressure: $\gamma_w H = 62.4 \times 18 = 1123.2\ psf$

Total pressure at bottom of footing = 1474.7 psf

Total active resultant = 0.5x1474.7x18 = 13,272 lb/ft

Answer is D

Solution 008

Initial effective stress at midheight of clay layer (17 ft below surface): $p'_1 = 5 \times 115 + 7 \times (125-62.4) + 5 \times (125-62.4) = 1326.2\ psf$

After lowering water table, effective stress: $p'_2 = 12 \times 115 + 5 \times (125-62.4) = 1693\ psf$

Ultimate consolidation settlement: $s = \frac{HC_c}{1+e_o}\log_{10}\frac{p'_2}{p'_1} = \frac{120\times 0.46}{1+0.45}\log_{10}\frac{1693}{1326.2} = 4.04\ in$

Time t = 3 months = 90 days; Drainage path H_d = 5 ft (double drainage)

From the settlement time relationship, the time factor is: $t = \frac{T_v H_d^2}{c_v} \Rightarrow T_v = \frac{c_v t}{H_d^2} = \frac{0.01\times 90}{5^2} = 0.036$

Corresponding degree of consolidation = 22%

Therefore, after 3 months, settlement = 0.22X4.04 = 0.89 in

Answer is A

Solution 009

There are 4 loading zones on the beam – the load function on them, left to right, are w = 0, w = constant, w = 0 and w = 0 respectively. As a result, the bending moment function is M = linear, quadratic, linear and linear respectively. This eliminates choices A, B and D.

Answer is C

Solution 010

The maximum compressive stress will occur at the upper left corner of the cross section, where the uniform compression P/A will combine with the bending stress components produced by the moment about either axis (Mc/I)

$$\sigma = \frac{20 \times 10^3}{0.2 \times 0.25} + \frac{20 \times 10^3 \times 0.07 \times 0.125}{\frac{1}{12} \times 0.2 \times 0.25^3} + \frac{20 \times 10^3 \times 0.04 \times 0.1}{\frac{1}{12} \times 0.25 \times 0.2^3} = 1.552 \times 10^6 \, Pa$$

Answer is B

Solution 011

With 2:1 side slopes and a depth of 5 ft, the width at the top surface = 20 + 2x10 = 40 ft
Area of flow, A = 150 ft²
Wetted perimeter, P = 20 + 2x5x√5 = 42.36 ft
Hydraulic radius, R_h = 150/42.36 = 3.54 ft

Velocity: $V = \frac{1.486}{0.015} \times 3.54^{2/3} \times \sqrt{0.008} = 20.58 \, fps$

Flow rate: Q = VA = 20.58X150 = 3087.2 cfs = 1995.2 MGD

Answer is C

Solution 012

Ultimate bearing capacity of a square footing, according to Terzaghi's theory, is given by
$$q_{ult} = 1.3cN_c + \gamma D N_q + 0.4\gamma B N_\gamma$$
From the supplied figure, for ϕ = 30°, N_c = 30, N_q = 18.5, N_γ = 22.5
$$q_{ult} = 1.3 \times 200 \times 30 + 120 \times 3 \times 18.5 + 0.4 \times 120 \times 5 \times 22.5 = 19,860 \, psf$$

Soil pressure at base of footing = column load + soil overburden = 140,000/25 + 120X3 = 5960 psf

FS = 19,860/5,960 = 3.33

Answer is A

Solution 013

Using a tributary width of 28 inches (2.33 ft) for each stud, the uniform load acting on each stud is 30x2.33 = 70 lb/ft

Maximum bending moment in stud: $M = \frac{wL^2}{8} = \frac{70 \times 10^2}{8} = 875 \; lb \cdot ft = 10{,}500 \; lb \cdot in$

Section modulus of stud (about major axis): $S = \frac{bh^2}{6} = \frac{1.5 \times 5.5^2}{6} = 7.56 \; in^3$

Maximum bending stress: $\sigma = \frac{M}{S} = \frac{10500}{7.56} = 1389 \; psi$

Answer is D

Solution 014

In the table below, cut and fill volumes between stations are calculated using the average end area method. The net volume is calculated in the last column (positive for cut and negative for fill). The cumulative earthwork volume is the sum of the numbers in the last column.

Station	Volume (yd³) CUT	Volume (yd³) FILL	Net Volume (yd³)
0 + 0.00			
	1032.4	370.0	+ 662.4
1 + 0.00			
	1340.7	141.3	+ 1199.4
2 + 0.00			
	1196.3	163.7	+ 1032.6
3 + 0.00			
	1445.7	560.9	+ 884.8
4 + 0.00			
			+ 3779.2

Answer is B

Solution 015

Maximum dry unit weight (Proctor) = 109 pcf
Therefore, required dry unit weight = 0.9x109 = 98.1 pcf
Weight of soil solids in embankment = 98.1x50,000 = 4.905x10⁶ lb
Dry unit weight of borrow soil = 120 ÷ 1.14 = 105.3 pcf
Therefore, volume of borrow soil needed = 4.905x10⁶ ÷ 105.3 = 46,598 ft³ = 1,725.8 yd³

Answer is B

Solution 016

The table below shows the precipitation in each of the three regions and the calculation of the weighted average as the average precipitation over the entire (268 acre) area

Region between	Area enclosed (acres)	Average Precipitation (inches)
1 & 2	147	0.4
2 & 3	76	0.75
3	45	0.9

The weighted average is the calculated as:

$$\bar{P} = \frac{\sum P_i A_i}{\sum A_i} = \frac{0.4 \times 147 + 0.75 \times 76 + 0.9 \times 45}{147 + 76 + 45} = 0.58 \, in$$

Answer is C

Solution 017

Using a Hazen Williams roughness coefficient C = 100 and a length L= 1 mile =5280 ft

$$h_f = \frac{4.725 Q_{cfs}^{1.85} L_{ft}}{C^{1.85} D_{ft}^{4.865}} = \frac{4.725 \times 12.5^{1.85} \times 5280}{100^{1.85} \times 2^{4.865}} = 18.3 \, ft$$

Answer is D

Solution 018

Volume of pond to be filled before it starts emptying = 0.78×760,000 = 592,800 gal = 79,251 ft³
At the rate of 2 cfs, this will require 39,626 seconds = 11 hours

Answer is C

Solution 019

Taking moments about support A,

$$\sum M_A = 20 \times 10 + 10 \times 14 + 40 \times 15 + 60 \times 30 + 40 \times 45 - 60 E_y = 0 \Rightarrow E_y = 75.67$$

Making a section through CD, DG and FG and then taking moments about G (see free body diagram below)

$$\sum M_G = 40 \times 15 - 75.67 \times 30 - \frac{4}{15.52} F_{CD} \times 15 - \frac{15}{15.52} F_{CD} \times 10 = 0 \Rightarrow F_{CD} = -123.43$$

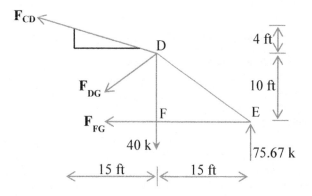

Answer is B

Solution 020

Factor of safety is given by:
$$FS = \frac{c}{\gamma H \cos^2 \beta \tan \beta} + \frac{\tan \phi}{\tan \beta}$$

For the riprap layer:
$$FS = 0 + \frac{\tan 36}{\tan 30} = 1.26$$

For the soil:
$$FS = \frac{600}{125 \times 20 \times \cos^2 30 \tan 30} + \frac{\tan 28}{\tan 30} = 0.55 + 0.92 = 1.47$$

The governing FS is 1.26

Answer is A

Solution 021

The fines fraction F_{200} = 28%. Therefore (since F_{200} < 50) the soil is predominantly coarse grained. First letter is S or G. Of the coarse fraction (72%), less than half (25%) is coarser than a no. 4 sieve. Therefore, the first letter is S. This eliminates A and D.

Since F_{200} > 12%, second letter of the classification is determined entirely by plasticity characteristics.
PI = 45 – 21 = 24. LL = 45 and PI = 24.
This plots above the A-line. So, second letter is C.

Answer is C

Solution 022

Mass of water = 1685 − 1418 = 267 g
Volume of soil solids, V_s = 1418 ÷ 2.65 = 535 cc
Therefore, volume of voids, V_v = 855 − 535 = 320 cc
Void ratio e = V_v/V_s = 320/535 = 0.6
Answer is D

Solution 023

Buckling about the weak axis is prevented because of the bracing provided by the nails.
The Euler buckling load (about the strong axis) is given by:
$$P_e = \frac{\pi^2 EI}{L^2} = \frac{\pi^2 \times 1.5 \times 10^6 \times \frac{1}{12} \times 1.5 \times 5.5^3}{(12 \times 12)^2} = 14848\ lb = 14.85\ kips$$
Answer is B

Solution 024

Using the reservoir surface as point 1 and open end of pipe as point 2, both of these points are at atmospheric pressure $p_1 = p_2 = p_{atm}$. Also, since the reservoir is 'large', by the continuity principle, $V_1 \approx 0$

Head loss (friction) in the pipe, using the Darcy-Weisbach equation, is
$$h_f = f\frac{L}{D}\frac{V^2}{2g} = 0.024 \times \frac{500}{2} \times \frac{V^2}{2 \times 32.2} = 0.0932 V^2$$
Writing Bernoulli's equation between points 1 and 2
$$\frac{p_{atm}}{\gamma} + 324.5 + 0 - 0.0932V^2 = \frac{p_{atm}}{\gamma} + 295.8 + \frac{V^2}{2 \times 32.2} \Rightarrow 0.109V^2 = 28.7 \Rightarrow V = 16.2\ fps$$
Flow rate: Q = VA = 50.98 cfs

Answer is A

Solution 025

Time of overland flow = $t_s + t_d$ = 5 + 13 = 18 minutes
Channel travel time = 1200 ÷ 5 = 240 seconds = 4 minutes
Time of concentration for point B = 18 + 4 = 22 min
For duration = 22 min and return period = 20 years, intensity = 1.5 in/hr

Answer is A

Solution 026

The tangent length (calculated from PI, PC pair – can also be calculated from PI-PT pair)

$$T = \sqrt{(1250-100)^2 + (320-817)^2} = 1252.8$$

The azimuth angle of the back tangent (calculated as inverse tan of departure divided by latitude)

$$Az_{BT} = \tan^{-1}\left(\frac{817-320}{100-1250}\right) = 156.63$$

The azimuth angle of the forward tangent (calculated as inverse tan of departure divided by latitude)

$$Az_{FT} = \tan^{-1}\left(\frac{900-817}{1350-100}\right) = 3.80$$

Therefore, the deflection angle I = 3.80 – 156.63 = – 152.83 (negative sign means deflecting left)

Since T = R tan (I/2), solving for R = 302.724 ft

Degree of curve D = 5729.578/R = 18.93 degrees

Answer is D

Solution 027

Height of warehouse = 26 ft
Perimeter = 2x(40+110) = 300 ft
Surface area of 4 walls = 300x26 = 7800 sq. ft.
Inclined length of roof = 2√($8^2 + 20^2$) = 43.08 ft
Surface area of roof = 110x43.08 = 4739 sq. ft.
Area of 4 openings to be subtracted = 4x8x13 = 416 sq. ft.
Total area of sheathing = 7800 + 4739 – 416 = 12,123 sq. ft.

Answer is D

Solution 028

Collapsing the 3D structures into a 2D one, and representing the *total* pad reaction on the left (2 pads) as R_L and the *total* pad reaction on the right (2 pads) as R_R, taking moments about R_L, we get:

$$9000 \times 14 + 800 \times 37 - R_R \times 36 + 8000 \times 56 = 0 \Rightarrow R_R = 16,767\ lb$$

With the load on the right, the maximum compression reaction on the ground will occur on the pads on the right.
Reaction on each pad = 8,384 lb

Soil pressure under the pads on the right (area = 7.07 ft^2) = 1,186 lb/ft^2

Answer is A

Solution 029

For concrete beams, the bending moment due to loading determines the extent of cracking experienced by the beam. The effective moment of inertia, used for calculating deflections according to the elastic theory, is between I_{cr}, the cracked moment of inertia and I_g, the gross moment of inertia. Only statement IV is correct.

Answer is B.

Solution 030

By using the 0.2% offset method (drawing a line parallel to the initial tangent through the strain offset = 0.2% = 2000 µε), the yield stress is 38 ksi

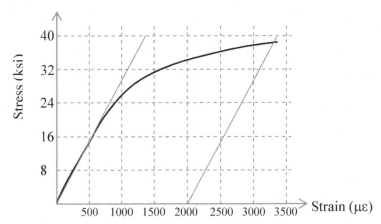

Answer is D

Solution 031

The thickness of sand that is affected by the pressure from the mat foundation is 15 – 3 = 12 ft = 180 in
Assuming that the mat exerts a uniform pressure of 600 psf on the entire layer,
Vertical strain = Vertical stress/E = $600/1 \times 10^6 = 6 \times 10^{-4}$
Vertical displacement = $180 \times 6 \times 10^{-4}$ = 0.108 in

Answer is A

Solution 032

II is incorrect. LS – EF – D = LS – (ES + D) – D = LS – ES – 2D = TF – 2D, which can equal TF only if duration D = 0 (which is a trivial solution)
Answer is B

Solution 033

In the long term, clay soils go through consolidation due to the expulsion of pore water. Results obtained from the CD (consolidated drained) triaxial test are good predictors of long term conditions. Therefore, statement I is correct.
The CD test involved allowing pore water to slowly drain from the soil and is therefore slower than the UU test. Statement II is correct and statement III is false. During the CD test, pore pressures are not allowed to build up. Statement IV is incorrect.
Answer is A

Solution 034
Material delivery = 540 yd³/hr (loose soil), which is equivalent to 540x0.83 = 448.2 yd³/hr compacted
Roller covers ground at 3 mph x 8 ft = 126720 ft²/hr. 0.5 ft thick layer gets compacted in 4 passes. Therefore each pass compact the equivalent of 0.125 ft, which means it compacts 15,840 ft³ (587 yd³) of soil per pass. This is ideal capacity. Working 50 minutes per hour, roller compacts 50/60x587 = 489 yd³/hr. Therefore, only 1 roller is needed to handle the delivery of the material.

Answer is D

Solution 035
Midspan deflection of a simply supported beam with point load is given by:
$$\Delta_{max} = \frac{PL^3}{48EI} = \frac{20 \times (24 \times 12)^3}{48 \times 29000 \times 1490} = 0.23 \text{ in}$$
Answer is A

Solution 036

The total vertical force (weight of concrete 3 & 4) + weight of soil (1) = 20,600 lb/ft
Horizontal friction force that can be mobilized under the footing = 0.6x20,600 = 12,360 lb/ft
$$FS = \frac{12,360}{5,400} = 2.29$$
Answer is C

Solution 037

The total volume of runoff that collects in the detention pond = 120 acre x 1 inch = 120 ac-in = 4.356x10⁵ ft³ = 12,333 m³ = 1.2x10⁷ L
Total mass of sediment = 6.167x10⁷ g = 6.167x10⁴ kg
Bulk specific gravity of the sediment = 80/62.4 = 1.282
Density of sediment = 1282 kg/m³
Volume occupied by sediment = 6.167x10⁴ kg ÷ 1282 kg/m³ = 48.1 m³ = 1699 ft³
Depth occupied by sediment = 1699 ÷ (2x43560) = 0.0195 ft = 0.234 inch
Answer is B

Solution 038

Breaking strain calculated as: Elongation at break ÷ Sample initial length

The values calculated for samples 1-5 are: 0.0728, 0.0703, 0.0719, 0.0716 and 0.0709

The average breaking strain = 0.0715 = 71,500 με

Answer is C

Solution 039

Depth ratio d/D = 30/48 = 0.625. For this depth ratio, for constant n, V/V_f = 1.0857

Longitudinal slope: $S = \frac{365.82-354.28}{1200} = 0.0096$

Velocity for pipe flowing full: $V_f = \frac{0.590}{n}D^{2/3}S^{1/2} = \frac{0.590}{0.014} \times 4^{2/3} \times 0.0096^{1/2} = 10.4\ cfs$

Actual velocity in pipe (when d = 30 in) is V = 11.3 fps

Answer is D

Solution 040

Underpinning (I) is commonly used to support existing structures to counter the possible loss of bearing support from adjacent excavation. Slurry walls (III) can be used to 'isolate' sensitive structures from construction activities.

Answer is B

SOLUTIONS TO WATER & ENVIRONMENTAL DEPTH EXAM
FOR THE
CIVIL PE EXAM

ANSWER KEY: WATER & ENVIRONMENTAL DEPTH EXAM

501	C
502	A
503	D
504	B
505	B
506	D
507	A
508	C

509	A
510	C
511	D
512	B
513	A
514	C
515	A
516	D

517	B
518	C
519	D
520	A
521	C
522	C
523	B
524	C

525	D
526	A
527	A
528	D
529	B
530	C
531	A
532	D

533	D
534	B
535	B
536	D
537	A
538	C
539	B
540	B

Solution 501

The TSS in the primary effluent = 0.4x400 = 160 mg/L

Assuming the primary sludge has specific gravity which is basically same as water, 5% solids (by weight) is equivalent to 50,000 mg/L

Assuming that the primary sludge flow rate is Q_{ps} and performing a mass balance at the primary clarifier:

$2.5 \times 400 = (2.5 - Q_{ps}) \times 160 + 50,000 \times Q_{ps} \Rightarrow Q_{ps} = 0.012$

0.012 MGD = 12,000 gpd

Answer is C

Solution 502

The total flow rate incident to the aeration tank = $Q_0 + Q_R$ = 2.5 + 0.5 = 3.0 MGD = 4.64 cfs

Hydraulic detention time: $t_d = \frac{V}{Q} = \frac{60,000}{4.64} = 12,931 \; seconds = 3.6 \; hrs$

Answer is A

Solution 503

Flow rate incident to the plant: Q_0 = 2.5 MGD

Volume of aeration tank = 60,000 ft³ = 0.45 MG

The food to microorganism ratio is given by:

$$F:M = \frac{S_o Q_o}{V_a X} = \frac{250 \times 2.5}{0.45 \times 1383} = 1.0 \; day^{-1}$$

Answer is D

Solution 504

Designating the stagnation point (at the head of the Pitot tube, where V = 0) as 1 and the free surface at the top of the liquid column as 2 and applying Bernoulli's equation (using the centerline of the pipe as datum, z = 0)

$$\frac{p_1}{\gamma} + 0 + 0 = \frac{0}{\gamma} + 0 + 22'' \Rightarrow p_1 = 62.4 \times \frac{22}{12} = 114.4 \, psf = 0.79 \, psi$$

Since the atmospheric pressure at 2 is zeroed, this value represents a gage pressure.

Answer is B

Solution 505

$$\frac{p_A}{\gamma} + \frac{V_A^2}{2g} + Z_A - h_f = \frac{0}{\gamma} + \frac{V_B^2}{2g} + Z_B \Rightarrow \frac{V_B^2}{2g} = 239.89 - 28.7 - 200.50 = 10.69 \, ft$$

$$V_B = 26.2 \, fps$$

Flow rate: Q = VA = 20.6 cfs = 9,249 gpm

Answer is B

Solution 506

Flow rate Q = 9000 gpm = 20.05 cfs; Velocity V = Q/A = 6.38 fps

Assuming the reservoir to be large enough that its surface velocity is negligible, the EGL elevation at A is:

$$E_A = \frac{p}{\gamma} + z + \frac{V^2}{2g} = 0 + 325.34 + 0 = 325.34 \, ft$$

At B, gage pressure = 60 psi = 8640 psf, and the EGL elevation is:

$$E_B = \frac{p}{\gamma} + z + \frac{V^2}{2g} = \frac{8640}{62.4} + 315.20 + \frac{6.38^2}{2 \times 32.2} = 454.29 \, ft$$

The total pump head = 454.29 − 325.34 + 45 = 173.95 ft

Pump power: $P = \frac{\gamma Q H}{\eta} = \frac{62.4 \times 20.05 \times 173.95}{0.88} = 247{,}310 \, lb \cdot \frac{ft}{sec} = 450 \, hp$

Answer is D

Solution 507

Head loss in circular pipes (using Darcy Weisbach method) is given by:

$$h_f = f \frac{L}{D} \frac{V^2}{2g} = f \frac{L}{D} \frac{Q^2}{2gA^2} = \frac{8fLQ^2}{g\pi^2 D^5} \approx \frac{fL_{ft} Q_{gpm}^2}{gD_{in}^5}$$

One can take either the path AB + BC + CE + EF or path AB + BD + DE + EF

Taking the path AB + BC + CE + EF, the head loss along these segments are 0.40, 2.15, 0.84 and 0.61 respectively. Therefore total head loss between A and F is 4.00 ft, which is equivalent to a pressure loss of 249.6 psf = 1.73 psi

Answer is A

Solution 508

Velocity at the entry (1) of the reducer: $V_1 = \frac{Q}{\frac{\pi}{4}d_1^2} = \frac{3}{\frac{\pi}{4}\times\left(\frac{1}{2}\right)^2} = 15.28\ fps$

Velocity at the exit (2) from the reducer: $V_2 = \frac{Q}{\frac{\pi}{4}d_2^2} = \frac{3}{\frac{\pi}{4}\times\left(\frac{1}{3}\right)^2} = 34.38\ fps$

Applying Bernoulli''s energy conservation principle between the entry (1) and exit (2) of the reducer

$$\frac{p_1}{\gamma} + \frac{V_1^2}{2g} + Z_1 = \frac{p_2}{\gamma} + \frac{V_2^2}{2g} + Z_2 \Rightarrow \frac{p_1 - p_2}{\gamma} = \frac{V_2^2 - V_1^2}{2g} = \frac{34.38^2 - 15.28^2}{2\times 32.2} = 14.73 \Rightarrow p_1 - p_2 = 14.73\gamma$$
$$= 14.73 \times 0.9 \times 62.4 = 827\ psf = 5.745\ psi$$

Answer is C

Solution 509

The friction head loss, according to Hazen Williams is:

$$h_f = \frac{10.429 Q_{gpm}^{1.85} L_{ft}}{C^{1.85} D_{in}^{4.865}}$$

The head loss per unit length: $\frac{h_f}{L} = \frac{10.429 Q_{gpm}^{1.85}}{C^{1.85} D_{in}^{4.865}} = \frac{10.429 \times 1500^{1.85}}{110^{1.85} \times 18^{4.865}} = 0.001$

Answer is A

Solution 510

In this problem, it is important to recognize that the flow is not necessarily at normal depth. Therefore, the values of Manning's n and longitudinal slope S are irrelevant.

With 2:1 side slopes and a depth of 4 ft, the width at the top surface = 12 + 2x8 = 28 ft
Area of flow, A = 80 ft²
Hydraulic depth, d_h = 80/28 = 2.86 ft
Velocity: $V = \frac{Q}{A} = \frac{405}{80} = 5.06\ fps$
Froude number: $Fr = \frac{V}{\sqrt{gd_h}} = \frac{5.06}{\sqrt{32.2 \times 2.86}} = 0.53$

Answer is C

Solution 511

Flow rate Q = 400,000 gpm = 200,000 ÷ 448.8 = 891.2 cfs
Instead of using a trial and error solution, we can use a flow parameter K, which is defined as

$$K = \frac{Qn}{kb^{8/3}S^{1/2}} = \frac{891.2 \times 0.015}{1.486 \times 20^{8/3} \times 0.008^{1/2}} = 0.0341$$

From tables for K (All-In-One Chapter 303 or CERM Chapter 19), for side slope parameter m = 2 and K = 0.0341, the depth ratio d/b = 0.1255
Therefore, the depth of flow, d = 0.1255x20 = 2.51 ft

Answer is D

Solution 512
Hydraulic jumps (I) and baffled outlets (V) are very effective as energy dissipation devices.
A sudden expansion (II) does cause some energy loss, but not a significant amount.
Floor roughening (III) does cause some energy loss due to friction, but not a significant amount.
Raising of the channel floor is not guaranteed to cause energy loss. That also depends on upstream conditions (subcritical or supercritical).

Answer is B

Solution 513
For flow to arrive from the most remote point in region A to point X = 45 min + 4700÷4 seconds = 64.6 min
For flow to arrive from the most remote point in region B to point X = 32 min + 3500÷4 seconds = 46.6 min
For flow to arrive from the most remote point in region C to point X = 56 min + 1700÷4 seconds = 63.1 min

Time of concentration for the entire watershed = 64.6 min

Answer is A

Solution 514
The upstream depth d_1 = 12 in = 1.0 ft

The conjugate depth d_2 is given by:

$$d_2 = -\frac{1}{2}d_1 + \sqrt{\frac{2V_1^2 d_1}{g} + \frac{d_1^2}{4}} = -\frac{1.0}{2} + \sqrt{\frac{2 \times 25^2 \times 1.0}{32.2} + \frac{1.0^2}{4}} = 5.75$$

By the continuity principle: $V_1 d_1 = V_2 d_2 \Rightarrow V_2 = \frac{V_2 d_2}{d_2} = \frac{25 \times 1.0}{5.75} = 4.35\ fps$

Upstream specific energy: $E_1 = d_1 + \frac{V_1^2}{2g} = 1.0 + \frac{25^2}{2 \times 32.2} = 10.7\ ft$

Downstream specific energy: $E_2 = d_2 + \frac{V_2^2}{2g} = 5.75 + \frac{4.35^2}{2 \times 32.2} = 6.04\ ft$

Percent loss of energy = 4.66/10.7x100% = 43.6%

Answer is C

Solution 515
The composite curve number for the entire 76 acre area: $\overline{CN} = \frac{\sum CN_i A_i}{\sum A_i} = \frac{74 \times 25 + 98 \times 5 + 80 \times 12 + 64 \times 34}{25 + 5 + 12 + 34} = 72$

Storage capacity: $S = \frac{1000}{\overline{CN}} - 10 = \frac{1000}{72} - 10 = 3.89$

Runoff depth: $S = \frac{(P_g - 0.2S)^2}{P_g + 0.8S} = \frac{(3.2 - 0.2 \times 3.89)^2}{3.2 + 0.8 \times 3.89} = 0.93\ in$

Answer is A

Solution 516

Converting each distance to miles and then calculating the product of Dia (in) x Length (miles), we have

Diameter (in)	Length (feet)	Length (mile)	DL (in-mi)
8	13,400	2.538	20.303
12	7,500	1.420	17.045
20	4,000	0.758	15.152
36	3,000	0.568	20.455
			72.955

Infiltration flow rate = 100x72.955 = 7296 gpd = 5.07 gpm = 0.011 cfs
Answer is D

Solution 517

In the continental United States, the U.S. SCS recognizes rainfall distribution types I, IA, II and III, as documented in the TR55 (Technical Release 55: Urban Hydrology for Small Watersheds, USDA, June 1986)

Answer is B

Solution 518

The first step is to separate the base flow in the stream hydrograph.

Time (hr)	Flow Rate (ft³/sec)	Net flow rate (ft³/sec)
0	20	0
1	50	30
2	70	50
3	120	100
4	60	40
5	20	0
	TOTAL	220

The area under this (Q vs t) curve represents the total volume of excess precipitation (runoff) and is calculated as:
$V = 3600 \times 220 = 792,000 \ ft^3$

Average depth of runoff: $d = \frac{V}{A} = \frac{792,000}{325 \times 43560} = 0.056 \ ft = 0.67 \ in$

Answer is C

Solution 519

The chlorine dose (concentration) is: 50 lb/day ÷ 8.34 ÷ 1.1 MGD = 5.45 mg/L
Residual chlorine = 0.5 mg/L
Therefore, chlorine demand = 5.45 – 0.5 = 4.95 mg/L

Answer is D

Solution 520

Digester volume = 100x40x12 = 48,000 ft³ = 0.36 MG
Calculating everything on a daily basis:
Solids quantity in digester = 20,000x8.34x0.36 = 60,048 lb
Solids quantity in decant (0.002 MG) = 400x8.34x0.002 = 6.67 lb
Solids quantity removed (0.004 MG) = 20,000x8.34x0.004 = 667 lb

$$SRT = \frac{Solids\ in\ digester}{Solids\ leaving} = \frac{60,048}{6.67 + 667} = 89.1\ days$$

Answer is A

Solution 521

Wastewater flow rate = 4000 gpm = 8.91 cfs
After dilution, the concentration of copper in the stream + wastewater mix is calculated as a weighted average:
$$\overline{Cu} = \frac{Cu_R Q_R + Cu_{WW} Q_{WW}}{Q_R + Q_{WW}} = \frac{0 \times 300 + 4 \times 8.91}{300 + 8.91} = 0.115 \frac{mg}{L} = 115 \frac{\mu g}{L}$$

In order to get this below 20 µg/L, the removal rate needs to be **at least**

$$\%R = \frac{115 - 20}{115} \times 100\% = 82.6\%$$

Answer is C

Solution 522

Assuming a residential wastewater flow rate per capita = 110 gpcd, the average flow rate from a population of 12,300 is
$$Q_{ave} = 110 \times 12300 = 1.35 \times 10^6 gpd$$

The peak factor, for a population of 12,300, can be calculated as: $\frac{18+\sqrt{P}}{4+\sqrt{P}} = \frac{18+\sqrt{12.3}}{4+\sqrt{12.3}} = 2.865$

Therefore, peak flow rate = 2.865x1.35 MGD = 3.87 MGD

Answer is C

Solution 523

For subsequent calculations, the elevation 108.78 is taken as the datum

Pumping rate Q = 800 gal/min = 1.783 cfs

At a radial distance r_1 = 1250 ft, there is no drawdown and therefore, elevation of piezometric surface h_1 = 205.98 – 108.78 = 97.2 ft

If h_2 = the piezometric surface elevation at the edge of the well (r_2 = 4.5 in = 0.375 ft)

$$Q = \frac{\pi K(h_1^2 - h_2^2)}{\ln\left(\frac{r_1}{r_2}\right)} \Rightarrow h_1^2 - h_2^2 = \frac{Q}{\pi K}\ln\left(\frac{r_1}{r_2}\right) = \frac{1.783}{\pi \times 0.003}\ln\left(\frac{1250}{0.375}\right) = 1543 \, ft^2$$

h_2 = 88.9 ft

Elevation of piezometric surface at the well = 108.78 + 88.91 = 197.69 ft

Answer is B

Solution 524

Total volume of runoff entering pond = 60 ac x 1 in = 60 ac-in = 217,800 ft³ = 1,629,000 gallons
Total volume of water lost due to evaporation in 1 hour = 2 gpd/ft² x (1/24) x 1.4 x 43560 = 122,000 gallons
Total volume of water lost through the weir = 1000 gpm x 60 min = 60,000 gallons

Net gain into the pond = 1,447,000 gallons = 193449 ft³

Elevation change (rise) = 193449 ÷ (1.4x43560) = 3.17 ft

Surface elevation after 1 hour = 233.17 ft

Answer is C

Solution 525

The total hardness is quantified as the sum of the ionic concentration of the polyvalent cations – Ca^{2+}, Mg^{2+} and Fe^{3+} in this case. Before combining them, they must be converted to the same units "mg/L as $CaCO_3$"

Ca^{2+}	15 mg/L x 2.5	37.5 mg/L as $CaCO_3$
Mg^{2+}	20 mg/L x 4.12	82.4 mg/L as $CaCO_3$
Fe^{3+}	8 mg/L x 2.69	21.5 mg/L as $CaCO_3$
Total hardness		141.4 mg/L as $CaCO_3$

Answer is D

Solution 526

The concentration of FSS in the wastewater = 40% of 380 mg/L = 152 mg/L

The concentration of solids removed = 80% of 152 mg/L = 121.6 mg/L

Quantity removed (lb/day) = 3x121.6x8.34 = 3042.4 lb/day

Answer is A

Solution 527
The limit for TDS in the secondary drinking water standards is 500 mg/L. Therefore, it is not in violation. This eliminates choices B and C.
For a single sample, the limit for turbidity in the primary drinking water standards is 1.0 NTU (as of Jan. 2002). **Therefore, the sample is in violation.**

The secondary drinking water standard for odor is 3.0 TON. Therefore, the sample is not in violation. This eliminates choice D

Answer is A

Note: The primary drinking water standard for nitrite is 1.0 mg/L (as nitrite). Converting the concentration of 3 mg/L to 3x14/46 = 0.88 mg/L as [NO_2-N]. Therefore, the sample is not in violation.

The primary (and secondary) drinking water standard for fluoride is 4.0 mg/L. Therefore, the sample is not in violation.

Solution 528
Average velocity: $V = \frac{Q}{A} = \frac{2514}{1323} = 1.9 \frac{ft}{sec}$

Reaeration coefficient: $k_r = \frac{3.3V}{H^{1.33}} = \frac{3.3 \times 1.9}{3.8^{1.33}} = 1.06$

Answer is D

Solution 529
K = 3 x 10^{-4} cm/sec = 1 x 10^{-5} ft/sec
For each side, N_f = 3, N_e = 7, H = 16 ft
Flow rate (per unit length) q = 1 x 10^{-5} x 3/7 x 16 = 6 x10^{-5} ft^3/s = 4.11x10^{-3} ft^3/min
Therefore, total seepage into the trench = 2 x 4.11 x10^{-3} ft^3/min/ft = 8.2 x 10^{-3} ft^3/min/ft

Answer is B

Solution 530

In the following table, the second column has the original unit hydrograph, while columns 3, 4 and 5 represent the scaled effect of hours 1, 2 and 3 respectively. For hours 2 and 3, the original data is shifted by 1 hour and 2 hours respectively. The last column has the sum of the highlighted columns.

Time (hr)	Flow Rate (ft³/sec/in)	Discharge (ft³/sec)			Total discharge (ft³/sec)
		due to 0:1 hr (0.5 in)	due to 1:2 hr (0.9 in)	due to 2:3 hr (0.3 in)	
0	0	0			0
1	80	40	0		40
2	180	90	72	0	162
3	320	160	162	24	346
4	210	105	288	54	447
5	0	0	189	96	285
			0	63	63
				0	0

Answer is C

Solution 531

The geometry of the circular section flowing partly full can be complicated. However, some tables can assist.
Depth ratio, d/D = 30/48 = 0.625, for which Q/Q_f = 0.7143 (for constant n)

Table 303.2 Velocity and Flow Ratios for a Circular Open Channel

d/D	P/D	A/D²	R/D	n/n_full	$n = n_{full}$		$n \ne n_{full}$	
					V/V_full	Q/Q_full	V/V_full	Q/Q_full
0.00	0.0006	0.0000	0.0000	1.0000	0.0000	0.0000	0.0000	0.0000
0.02	0.2838	0.0037	0.0132	1.1118	0.1407	0.0007	0.1266	0.0006
0.04	0.4027	0.0105	0.0262	1.1548	0.2219	0.0030	0.1922	0.0026
0.06	0.4949	0.0192	0.0389	1.1847	0.2889	0.0071	0.2439	0.0060
0.58	1.7315	0.4724	0.2728	1.2250	1.0591	0.6374	0.8646	0.5203
0.60	1.7722	0.4920	0.2776	1.2172	1.0716	0.6718	0.8804	0.5519
0.62	1.8132	0.5115	0.2821	1.2090	1.0831	0.7059	0.8959	0.5839
0.64	1.8546	0.5308	0.2862	1.2005	1.0936	0.7396	0.9110	0.6161
0.66	1.8965	0.5499	0.2900	1.1916	1.1031	0.7729	0.9257	0.6486

Flow rate (pipe flowing full) is given by:

$$Q_f = \frac{0.463}{n} D^{8/3} S^{1/2} = \frac{0.463}{0.015} \times 4^{8/3} \times 0.01^{1/2} = 124.4 \, cfs = 80.4 \, MGD$$

Therefore, flow rate Q = 0.7143 x 80.4 = 57.4 MGD
Answer is A

Solution 532

Q = 3 MGD

TSS = 1,100 mg/L
Suspended solids load = 3 x 1,100 x 8.34 = 27,522 lb-TSS/day
Maximum solids load on each filter = 12 lb-TSS/ft^2-day
Total filter area required = 27522 ÷ 12 = 2,294 ft^2 = 10.2 filters. Therefore, 11 filters are needed based on capacity.
To accommodate the provision of filter washing, number of filters provided = 13

Answer is D

Solution 533

The intensity during each period should be calculated from the incremental depth of precipitation during that period.

Time interval (min)	Depth (inches)
0 – 10	0.2
10 – 20	0.3
20 – 30	0.4
30 – 40	0.6
40 – 50	0.3
50 – 60	0.2

Peak intensity occurs during 30 – 40 min interval.
Intensity = 0.6 in/10 min = 3.6 in/hr

Answer is D

Solution 534

Head difference = 945 – 905 = 40 ft
Length of seepage path = 160 ft
Hydraulic gradient = 40/160 = 0.25
Area of flow = 120 x 6.5 = 780 sq. ft
Permeability k = 9.5 ft/day = 0.0066 ft/min
Discharge Q = KIA = 0.0066 x 0.25 x 780 = 1.286 ft^3/min = 9.62 gal/min

Answer is B

Solution 535

The 5-day BOD is calculated as: $BOD_5 = \dfrac{DO_i - DO_f}{\frac{V_{sample}}{V_{total}}} = \dfrac{6.3 - 2.5}{15/300} = 76 \; mg/L$

Ultimate BOD is calculated as: $BOD_{ult} = \frac{BOD_5}{1-10^{-kt}} = \frac{76}{1-10^{-0.1 \times 5}} = 111.1 \ mg/L$

Answer is B

Solution 536

$$DWEL = \frac{R_f D \times Body\ weight}{Consumption} = \frac{0.06 \times 70}{2} = 2.1 \frac{mg}{L}$$

Answer is D

Solution 537

Flow rate Q = 1.5 MGD = 2.32 cfs
Tank volume: $V = Qt = 2.32 \times 120 = 278.5 \ ft^3$
At a temperature of 50°F, viscosity of water, µ = 1.41x10^{-5} lb-s/ft^2
Power P = 35 hp = 35x550 = 19,250 lb-ft/sec
Power: $P = G^2 \mu V \Rightarrow G = \sqrt{\frac{P}{\mu V}} = \sqrt{\frac{19250}{1.41 \times 10^{-5} \times 278.5}} = 2,214 \ sec^{-1}$

Answer is A

Solution 538

Bioaccumulation is defined as the accumulation of chemicals *in the tissue of organisms* through any route, including respiration, ingestion, or direct contact with contaminated water, sediment, and pore water in the sediment.
Algal blooms, especially if the algae are of the toxic variety, *can lead to* bioaccumulation, but are not an example of bioaccumulation themselves. Buildup of greenhouse gases is an example of a buildup of potentially harmful gases in the atmosphere, which is part of a biosystem, but does not directly guarantee an uptake by organisms in that system. Similarly, sediment deposited in a detention pond may contain toxins that are then taken up by organisms, but the phenomenon of sedimentation is not bioaccumulation.

Answer is C

Solution 539

Potassium permanganate (KMnO$_4$) is commonly used to remove odors such as "rotten egg" smell from water.

Answer is B

Solution 540

At the rate of return (i), the present worth should be zero.
The $350k capital expenditure is a present value (P).   NEGATIVE
The $25k reduction in annual costs is an annuity (A)   POSITIVE
The $200k increase in salvage value is a future sum (F) POSITIVE

Converting all of these to present worth, the net present worth can be written:

$$PW = -350 + 25\left(\frac{P}{A}, i, 20\ yrs\right) + 200\left(\frac{P}{F}, i, 20\ yrs\right) = 0$$

For i = 5%, PW = 37k
For i = 6%, PW = -0.9k   Actual answer 5.97%
For i = 7%, PW = -33.5k

Answer is B

Made in the USA
Las Vegas, NV
20 March 2024

87506216R00044